Gladys

The Rescue Crayfish

From the Animal Rescue Series

by Terry Blackburne

Illustrations by Kathy Rohn

Gladys the Rescue Crayfish
From the Animal Rescue Series

ISBN 978-1-7331017-5-2

This book is dedicated to all of the animals that are studied in science classrooms.

Gladys wants everyone to know that animals who are used in science labs deserve to be treated with kindness and respect.

She also wants people to adopt animals after helping children learn about them in the science classroom and give the animals a happy life.

Other books in the *Animal Rescue Series*
by Terry Blackburne:

Zeus the Rescue Cat
Benny the Rescue Dog
Raphael the Rescue Turtle
Ben the Rescue Horse
Goldzilla the Rescue Goldfish
Freedom the Rescue Hawk

Copies can be ordered on Amazon.com or other online retailers. To place a bulk order directly from the author contact tmblackburne@gmail.com.

Contents

CHAPTER 1

Louisiana Bayou

I lived most of my life without a name because my species was not usually lucky enough to be a pet. I was a crayfish, like a mini lobster, a bottom feeder, who was considered a "cockroach of the sea". Humans typically ate my kind as part of a meal and crayfish were prey for lots of other animals too. People often used my relatives as bait on a fishing hook thrown in the water to catch a fish. As none of those things were my fate, I can't be too upset that I received my name, Gladys, later in my life when I ended up as someone's pet crayfish in a big tank with lots to eat.

My birth took place on a dark summer night in a damp Louisiana bayou. The warm water encouraged me to peel off the egg that enclosed me. I first viewed the underside of my mother's tail to which I was attached. Her large, segmented tail fanned out to accommodate me and my siblings. There were about forty of us, little as pinheads, holding tightly onto my mom's spinnerets.

Louisiana Bayou

We floated in muddy water within a cluster of sticks and grass. Sounds of crickets and owls echoed from everywhere. With my mother's crusty body to protect me, I felt safe as she curled up her tail so my siblings and I had extra protection. I didn't want to become dinner for any of the avian creatures of the night as they circled our little nesting place.

The next morning, I discovered even more animals that appeared with the sunrise. Birds of all shapes and colors, chirping and squawking, flew from trees to bushes to branches to logs in the water. If they dove into the water looking for food, my mother closed her tail tightly over us and bored into the muddy bottom of the stream.

Muskrats, nutrias and other rodents drank water at the edge of the bayou, searching for food among the leaves and branches on the shoreline. Their front paws with long fingernails scratched at the substrate. I'm glad they didn't reach too deep into the water for something to eat because their large, yellow front teeth scared me.

Alligators with open jaws and rows of teeth waited for some movement in the bayou to pounce on their prey. I clung to my mother's tail with all my might. I didn't want to be someone's breakfast before I was a day old. I don't think the predators saw me at all since I was so tiny, but my mother had to be careful and she slid deeper into the mud at the bottom of the water.

I learned to eat by venturing out from the protective underside of my mother and tasting many different oddities within my swimming range. I liked the six-legged bugs that tasted of honey and butter and hated the green slime, called algae, which grew liberally on the rocks and logs. The wiggly worms and eight-legged spiders were especially delicious. After every trip away from my mother, my siblings and I swam back to her protective tail and cuddled up as she curled us close to her body.

Each day I traveled further and further from my mom. She encouraged me to curl my tail under me to go backwards and showed me how to walk on the muddy bottom with my plethora of legs. Swimming with my tail was my favorite activity because each time my back end swooshed under me, it created a little wave of water and a scratching sound on the bottom. Several of these motions sent me many yards in the opposite direction. It was such fun!

I liked stretching my many legs as they grew. One day I noticed that some of them were developing differently than my many back legs. My two front ones grew quickly and became much larger, longer and pointier day by day. I opened and closed them easily, cutting things in half like long worms or big bugs. When a big bird tried to peck at me, I snapped them shut to scare the bird away. What a great protective feature! I practiced using my claws on other animals and plants and delighted in their ability to help me get food and frighten animals away.

I also grew an exoskeleton all over every part of my body except my eyes. This crunchy protective outer shell was just like my mom's and made me feel safe in my own body. By this time, I was too big for the underside of my mother but still stayed near her and the other baby crayfish like myself. I loved my home in the bayou.

One day, I discovered two very long things sticking out from above my eyes. Called antennae, I smelled and felt things with them. This was fantastic! I searched for food with these long appendages. When I found something to eat I walked with my spinnerets, each with a tiny claw on its end, over to the food and ate it. My mouth opened on the underside of my body, near where a neck would be if I had one. My body consisted of three parts, a head, an abdomen and a cephalothorax. People called me a crustacean, just like shrimp and lobsters. Stalks that stuck up from my exoskeleton, held my eyes. Tall and free from my body, they moved in all directions seeing 360 degrees around me. My eyesight was not so great, though, so I used my antennae more often to know what was going on around me.

Life in the bayou was pretty special for crayfish if we survived being eaten by predators. There was an endless supply of mud, twigs and sticks to hide under. When I hid there, the birds didn't see me from the sky. Sometimes the little animals with long claws like the rats and nutrias, dug down in the mud looking for me and my crayfish friends. That's when I'd swim away by curling my tail in rapid succession, sending me in a backwards swim. This propelled me away from their busy hands and hungry mouths.

I knew that I swam faster than most of the other crayfish in my bayou. I enjoyed practicing my swimming every day over and over until my tail muscles were strong. I exercised my spinnerets and they also developed muscular mass. Because of my strong ability to swim, I wasn't caught by any predators.

Each day my siblings and I played in the mud and leaves chasing each other and hiding from each other. It was a lot like hide and seek that children play on land. Sometimes I hid under a rock but my brothers always found me there. The best hiding place was among the roots of trees that grew into the water. Green, slimy moss thrived on the bark and was the perfect place to hide. My sister got wise to me, though, and discovered my favorite hiding place after a few days. Then I scurried to find another equally excellent hideout. When I was *"it"*, my brothers and sisters helped me to discover other unusual quiet spots like the inside of a rotting log. I loved my little crayfish life. We all slept together with my mom among the decomposing leaves of the bayou.

CHAPTER 2

Captured

One dreary, rainy day, two boys came to the shore of our bayou with a net on a long pole. They had a bucket with them and a cover to put on top of the bucket. The boys took off their shoes and waded into the warm water of our home. "What are they doing?" I thought. I never saw a human come into our water before. What a strange, new expereience!

They dipped the net into the pond; it was strong and the pole was sturdy. The young men started to scoop up water and leaves with their tool. I hid under the mud assuming I was safe. That's when the unthinkable happened. They dug right into the mud and pulled me and several of my fellow playmates up out of the water. I scooted deeper into the mud but discovered that it was in the net that lifted higher and higher out of the water. Then, with a loud *PLOP*, my captors dumped the mud into a bucket and screwed the top on tightly. I was caught. What did this mean for me? Cowering in the mud at the side of the bucket, I tried to remember what my mom taught me about getting away from predators. I flailed my front claws opening and closing them threateningly. But who was I trying to scare? There didn't seem to be an identifiable predator around, just a plastic bucket.

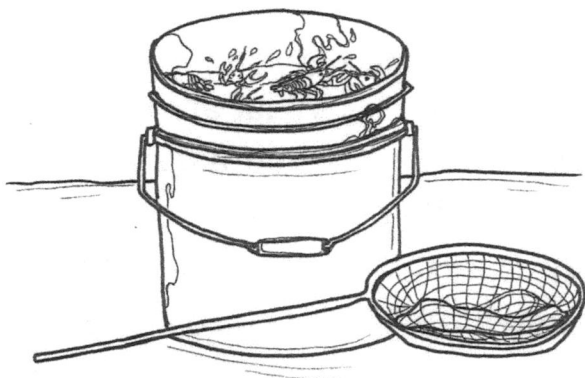

Captured

The boys picked up the pail and started to walk, swaying me and the other crayfish in an unfamiliar motion. It wasn't unpleasant but I didn't know where I was going. As the bucket swung rhythmically, I searched in the mud for my mother. She escaped the capture and something told me I would never see her again. Sadness overwhelmed me. The fight to survive had tired me out and I dug deeper into the cooling mud and fell asleep.

When I awoke, I was in a bright room with many tanks filled with water. There were hundreds of crayfish like myself in the tanks. There was a man in the room who told the boys, "Good job guys, these are some nice looking crayfish. They will be great to study in the science classrooms."

"What's a science classroom?" I thought.

Over the next few days, I lived in a glass tank with lots of water. There were fake logs and plastic tunnels to hide in but with hundreds of crayfish all stuffed into a confined space, many fights broke out. Some crayfish lost one or both of their front claws during these arguments. I knew from my mother's teachings that our front claws grow back over time. Once, my little sister's claw was severed by a falling rock and it slowly reappeared over a few weeks. I didn't want to lose my claws, though. I thought it might hurt to lose an appendage and I didn't want to be clawless for weeks. So, I found a nice log to rest in and didn't fight with anyone. Everyone else considered me a loner.

Twice a day, the man in charge put several pellets that tasted like shrimp into our container. After just one feeding, I realized that although most of them floated, some fell to the bottom. Instead of shoving for the food on the top, I scooted to the bottom and grabbed the pellets that sank. In this way I avoided competing for my nutrition.

After a few days, a group of crayfish, including me, were put in a box with wet leafy greens in the bottom and holes in the top. I panicked because I couldn't see anything. Never had I been in such close quarters with so many other crayfish. I wondered if I would ever see daylight again. I worried that I would die in this tight space without any water or food. Some of the crayfish started fighting with each other. Crying, I found a corner of the box and patiently waited.

It wasn't long before I felt our box being moved faster than I had ever been able to go on my own. I heard a loud noise from the motor of a truck. The trip was short and our box was carried into a building with lots of shouting and giggling sounds. A bell rang and our enclosure was carried through a door into a warm space. I still couldn't see anything but my box stopped moving and settled on a flat, solid surface.

CHAPTER 3

Intro to a Science Classroom

When my box was finally opened up, I looked out on a sea of little human faces smiling and jumping up and down with sheer delight. There were more tanks filled with water in this room too. "Oh, no," I thought. "Not another crowded tank."

"Just one crayfish in each tank," the teacher, named Terry, said. "Remember, they need their own space and the boy crayfish often fight with each other if the tank is not big enough."

No one was trying to fight with me. "I must be a girl," I discovered.

As the other crayfish were being lifted out of the box, some of them tried to bite the teacher with their large front claws. They flapped their tails and snapped their claws open and closed trying to get free. One of them pinched the teacher's hand with his largest claw. "Ow!" yelled Terry. "That hurts! Students, try not to put your fingers or hands right in front of the crayfish. They are scared and just trying to protect themselves. But it hurts, nevertheless."

I didn't fight Terry at all. I got a feeling that she was going to be kind to me. When she picked me up, she had a soft, gentle touch. Then she put me in a small, one gallon plastic tank that had little colored rocks on the bottom and lots of plants floating on the top. I saw moneywort, anacharis, duckweed and salvinia. I remembered eating all of these plants in the bayou so I knew I wouldn't go hungry. But there was nothing on the bottom of the tank and it felt lonely.

Then, the teacher said, "Children, remember the castles we bought at the pet store? Let's rinse them off with clean water and put them in the tanks for the crayfish to hide in." And so I was given a bright pink and yellow castle of my own with two doors and two turrets. I walked through one of the doors and was enclosed in a cool, dark place.

When I poked my antennae out of one of the doors, I discovered that one of the students had put some tiny worms, called bloodworms, into my tank. Red and juicy, I picked some up with my swimmerets and put them in my mouth. Yum! I never tasted anything so delicious in all my life. The bloodworms were soft and rubbery and fit perfectly into my small

16

mouth. They exploded with a taste of warm, flowery sunshine and I decided that from then on, bloodworms were my favorite food.

Four little faces peered down at me. They looked very different from me so I stared back. Each child had two eyes but they were recessed into their heads instead of on stalks like mine. How weird! I guessed that they couldn't see around them as well as I could. I was able to look 360 degrees around me in a full circle. The children had to move their entire head to look around them. There were eyelids that constantly opened and closed over their eyes. How annoying that must be. I had no such impediment to my eyesight.

They also had hair on top of their heads where my head was smooth with my exoskeleton. I wondered why they grew the strange hair and what advantage that might give them. I liked my smooth head a lot better. It was very strange to see little flaps on the sides of their heads where my ear holes are. Maybe those protrusions helped them to hear but I couldn't imagine how. Right in the middle of their face was a small mountain with two holes in it. I couldn't figure out what that was for since I don't have anything even remotely similar.

I felt sorry for these children because they only had two arms and I had more than twelve appendages including my pincers, legs and swimmerets. On the tips of each of their arms they had claws that had five parts compared to my two part claws. Their heads were not part of their abdomen but rather had a thick stick that connected it to their torso. Why did someone need that? My body seemed to be so much more streamlined. I supposed that was why I lived in the water and they didn't.

I had to strain to see what they looked like below the table but I was able to see a bit. It looked like they had two more appendages but they were covered with different colored materials that didn't allow me to see what was underneath. How uncomfortable that must be to have your body covered up like that. Oh well. I'm glad I'm a crayfish and not a human. I wouldn't want to put that weird stuff all over my body. I'm glad I am naked and free to enjoy my watery existence.

The students also made a lot of noise with a very large opening in their face. They called to each other across my tank and made loud sounds that reminded me of birds squawking as they pointed at me. It was interesting how they communicated with each other with sounds rather than bodily movements like crayfish. We moved our antennae and touched each other to let another crayfish know what we wanted. These little people were truly very strange.

I found out that this was a science classroom and that these four children were going to learn everything they could about me. "OK," I thought. "If you feed me more blood worms, I will let you examine and study me."

CHAPTER 4

Scientific Studies

For the next few weeks the children came to our classroom every morning and I got to show off. I didn't know where the little ones were the rest of the day and night but when they came to the classroom, I was the center of attention and I liked it.

The first day, they all drew a picture of me in their notebooks and labeled all of my parts, even my spinnerets. It took them a long time to count them and two of them even stuck their hands into my tank to move me around while they counted them. I was glad they didn't hurt my delicate antennae which they drew with accurate precision. When they discovered that I had two pairs of antennae - a short pair and a long pair, they questioned why I had them. I didn't know how to communicate to them that my short pair tasted the water and tested food before I ate it while my longer pair felt things around me. I remained still and didn't bite them with my pincers which they called my front claws. They discovered that in addition to my two large front claws, I had little claws on the tip of each of my four pairs of legs.

Next, they drew my cephalothorax and abdomen. With much excitement, they realized that I was really one long connected body. "Our crayfish has no neck," one of the children shouted.

"You are right," said the teacher. "Is that like any other animal that we have studied this year? What other type of animal has a head, thorax and abdomen?" she said.

"Insects!" they all said at the same time.

"Yes! Crayfish are crustaceans, though." Terry explained. "However, they are often referred to as the cockroaches of the sea because they look like cockroaches and eat leftover food from other animals. Actually," the teacher continued, "insects and crustaceans are all arthropods because they both have segmented bodies with paired, jointed appendages. The name arthropod comes from the Greek word for jointed feet."

On the second day, the students discovered the presence of my exoskeleton. "Feel the outside of the crayfish's body," one of them posed. "It's kind of crusty and a little bit hard."

"Let's get out our laptops and find out what it's called, " said the instructor. The children had funny looking shiny things that opened up like a book without pages. They poked their fingers along the bottom of it and images appeared on a screen. They called them computer laptops. What a surprise when a picture of me came up on the screen with lines pointing to different parts of my body! My exoskeleton was labeled and described in ultimate detail. They learned that my body was covered in a semi-hard protective covering and that I had no bones. Thus, I belonged to the largest group of animals in the world called invertebrates, animals with no backbone. The computer said that my exoskeleton is a supportive outer covering that all arthropods have. In order to grow, I must shed this exoskeleton, called molting, and develop a bigger one every few weeks. I realized it already happened to me. On the day I arrived in the classroom. I actually crawled out of my outer covering and was soft for a few hours until my body hardened again. I could see a perfect outline of my body, including my legs, spinnerets and antennae drifting in the water next to me. Upon closer inspection, it appeared to be hollow, just my shedded exoskeleton. I snacked on it and the rest disintegrated into the water before the students even saw it.

The third day, they took me out of the water. There was a big fuss about who would touch me and try to lift me out of my watery home.

"He's slimy," one of the girls said.

"I don't want to get bitten," the other girl cried. Finally, one of the boys grabbed my antenna and tried to pick me up. Immediately, my tail whipped under me and I moved backwards with a huge swish of water. I put my claws up in front of me and opened and closed them menacingly. Perched up on my spinnerets and waving my front claws, I resembled a marionette except without the strings. I feared being dropped onto the floor. All four students stepped back in deep surprise.

"Let me help you," Terry exclaimed. She came over and pinched me gently on either side of my abdomen and pulled me out of the water. I couldn't get my claws back that far to bite her but I realized that her grip was tight and sure. She held me upside down and even though she held me securely, I was nervous about falling into the water on my back and not being able to flip around to right myself. Balancing, I stayed very still to help her and wished I could close my eyes but I didn't have eyelids on my stalk eyeballs.

They all looked at my underside and confirmed that I was a girl crayfish. "This crayfish does not have an extra pair of swimmerets so she is a girl," the teacher said. One of the little girls asked why I was missing these swimmerets but I didn't understand what Terry told her because I was worried about protecting my soft underside.

Then she picked up another crayfish from the next table and held him the same way, upside down, right next to me. We looked imploringly at each other hoping to hatch a plan to save ourselves. No such luck. "See how this boy crayfish has one extra pair of legs under him. That means he is a boy."

"Wow," I thought. "Am I missing a pair of appendages?" It didn't feel like I was missing anything. I felt perfectly complete. "Oh Well. I'm glad I'm a girl without even more legs to worry about," I convinced myself.

The fourth day, the students picked me out of my tank with a net attached to a long handle. They put me in a little plastic dish and tried to feed me all sorts of weird things.

A meal worm came first and I gobbled it up right away. Next I was offered an orange stick that was hard and inedible so I just poked at it with my antennae. I wouldn't go near the big, white oval that floated on the top of the container. It smelled good but it got soggy immediately as it absorbed the water. However, I tasted the red round thing they put in my dish. The children called it a strawberry. I had to break off pieces with my claws so they were tiny enough to fit in my mouth. It tasted sweet and juicy. Yum!

As they studied me, the students noticed that my mouth was not located where they expected it to be. "Look where our crayfish is putting the food to eat it. It looks like she is pulling the food to the underside of her neck or where there would be a neck if all her body parts weren't connected," they exclaimed.

Then the children held little round instruments over one of their eyes and it made that eye look tremendous and bulbous.

"I see her mouth," one of the girls yelled. "She has two yellow teeth that open and close as she ingests her food."

"Those are called mandibles," said Terry, proud that her students discovered something about me without her help. I looked closer at them and noticed that the things in their mouths that they use to chew food were white and numerous. How strange! I prefer my two yellow mandibles. They made a chart in their notebooks and put a check mark next to all the things I ate and an X next to all the things I wouldn't eat.

The fifth day, the class was instructed to discover how I moved from place to place. To encourage me to travel in my tank, the students put food pellets clear on the other side so I would have to go get them. I walked slowly along the bottom of the water with my swimmerets to find my lunch.

Then one of the children dropped a rock into my tank which scared me so I whipped my tail underneath me in a fan motion and swam backwards away from the rock, fast as lightning. The students laughed and wrote notes in their journals. They drew pictures of my tail and tried to imitate my movements running around the classroom on their two legs.

"How sad," I thought. "These little humans didn't seem to have any tails."

It continued like this for a few weeks, each day the students were given an assignment and I would cooperate and help them learn about me. I felt very special.

When it seemed like they had learned everything they could about crayfish, the four little kids at my table made a giant poster and put all their work on it for everyone to see. They displayed pictures of me along with a labeled diagram. They had explanations of my different body parts like my antennae, cephalothorax and abdomen, pincers, legs and swimmerets. They drew a diagram of my food plates and listed all the food I liked to eat. There were charts about my behavior and eating habits. They even wrote a story about when I met the crayfish at the next table and ran into my castle to hide.

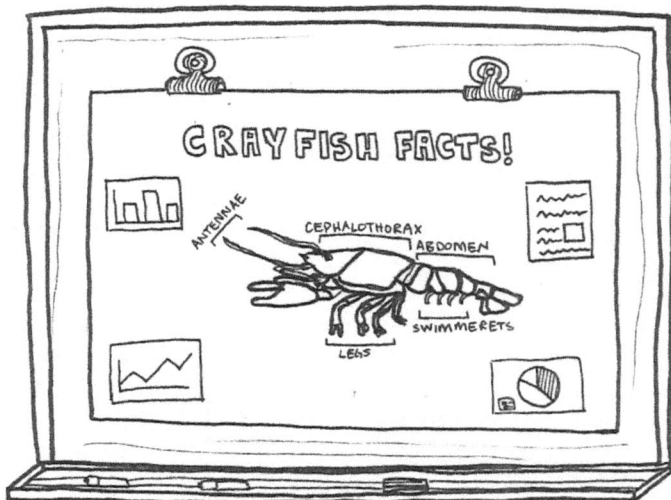

Their parents came into the school one night to see the reports and meet me. They were all dressed up and brought snacks for the children to eat. When they peered into my tank, I straightened my antennae up, held my two front claws perfectly in front of my body, and fanned out my tail. I wanted the children to be proud of me while their parents were in our classroom.

"She is very pretty," one of the adults said about me. "You must have had a lot of fun with your crayfish."

"We did, we did!" the children shouted in unison.

CHAPTER 5

Fake Pond

This, however, turned out to be the end of my adventures in the science classroom and with my wonderful little friends. My water was starting to get dirty and smelled putrid because there was no filter in my small space. The teacher instructed the children to put their crayfish in a big box and then clean the tanks in the classroom sinks.

My group of students, familiar with holding me by now, lifted me out of my tank and each one held me in the palms of their little hands.

"Goodbye, crayfish. We love you," the students said as they blew kisses to me and smiled.

I got excited. Maybe they were going to return me to the bayou. I longed to be free again in my wonderful, natural ecosystem. I wanted to see my mother and siblings again. Wouldn't that be a nice reward for a little crayfish who helped these humans learn about a crustacean like me? I hoped with all my might.

As I scrambled into the box of ten crayfish, I looked at them to see if the anticipation of freedom had crossed their minds as well. They looked dejected and not hopeful like me. Most of them curled their tails under them in a protective stance that said they were nervous and stressed out. I reached my claws high above me imploring the teacher to be kind to us and reward us for all of our help.

The classroom smelled really bad as all the tanks were dumped into the sinks. I knew it was because of me and the other crayfish. We had to poop somewhere and we rarely got out of our enclosures. That's why the water started getting toxic.

In the wild, miles of water filled our bayou and several bottom feeders, like catfish and snails, cleaned up after us. Also, benthic, worm-like animals decomposed waste materials. None of these decomposers were in our classroom aquarium. I was glad that I didn't have to stay in my dirty tank anymore. But where was I going now? I was worried.

At the end of the day, the teacher carried us in the box to her car and we rode home with her. She carried our group of crayfish to a large tank of water in her grassy backyard. Three hundred gallons of water glistened in a horse trough. It looked like a small pond. There were tunnels to hide in and rocks to climb around and hide under. There was even a big filter that sprinkled clean droplets across the surface of the water.

She dumped our box of ten crayfish into the trough. We were in the sun and I thought this might be a nice place to live. I swam around and then settled on the bottom of the tank.

Just then, the enormous jaws of a turtle snapped right in front of my face. His mouth had no teeth but his enormous beak had a sharp point on it and I could see his little pink tongue. He was hungry, there was no doubt about that, and I looked like a delicious meal to him. I watched one of my crayfish friends get eaten by another turtle, exoskeleton and all. In one big gulp, the turtle opened his mouth and swallowed my fellow crayfish.

"Oh, no, predators," I thought. "Am I back in a mini bayou?" But the only things here are turtles - no birds, rodents or muskrats. "I have to find protection." I panicked.

I swam as fast as my tail would take me and hid in one of the tunnels. "Chomp!" came the jaws of a different turtle who found me in there. I climbed onto one of the rocks jutting out of the water. "Waroomp!" came the beak of another turtle.

"Oh, no," I thought. "The turtles can fit into these tunnels and climb on the rocks. I need to find another place to hide." There was no mud on the bottom of this fake pond, there was only a square thing with a grate on top of it sucking in water and spouting it out the other end. It looked strangely inviting because the turtles clearly stayed away from it. The filter had a powerful force somewhat like gravity pulling things toward it. It had a softly swishing sound and looked unnatural like something totally manmade.

"Maybe I could get myself sucked into that filter." I thought. "I don't think the turtles can fit in there." I swished my tail and swam near the thing. Nothing happened. I touched the top of the filter with one of my front claws but I didn't feel any suction. Finally, I sat right on top of the grate and ZAP I got taken into the filter which sucked in my antennae so hard that I thought they would tear off. Next came my cephalothorax and abdomen and lastly my pincers that were pinned onto the sides of my body rather than extending in front of me for protection. "Oh well!" I thought. "This is what I wanted. This filter had better protect me from those turtles after the uncomfortable sensation it caused me to get in here."

A soft spongy thing inside the apparatus cushioned me. Lots of bloodworms squiggled and wiggled around and it looked like they were dancing to some silent music playing. I ate as many worms as I could until I felt full. "I guess this isn't such a bad place to hang out for a while," I thought. I stayed there for many days just enjoying the unlimited food, protective covering and quiet swooshes of the filter water until Terry came to clean the filter one day.

Once a week in the summer months, Terry pulled the filter out, hosed off all the dirt and waste with clean water, and put it back into the tank. The water in the big trough stayed clean because as dirty water went through the filter's grate, the spongy part of the filter trapped all the dirt and debris and then clean water was expelled out the other end. Hosing off the filter each week kept the tank clean.

When Terry took the filter apart to hose it off, she saw me hiding with my head down and my tail curled under. I was nervous because I didn't want to lose my protective home and face those hungry turtles again. Even though it was dark in the filter, I feared being put into the big waters again.

"Oh, my!" she said when she spied me. "You are a very resourceful crayfish. You survived the turtles eating you. What a great place to hide!" She cradled me in her hands and looked me over. "You deserve to live a long and happy life free of any predators or science experiments. I know just the place for you."

CHAPTER 6

Pirate Ship

Terry carried me into her warm house and dropped me gently into a huge glass tank of water filled with little fish, snails, castles and a pirate ship. There was a filter at the top of the tank spewing clean water into the tank and several plants floating on top of the water. What a beautiful environment. I cautiously scanned my new home and discovered no turtles. Thank goodness!

I looked around slowly. There were no other crayfish either. Checking to see what type of fish and plants lived there, I recognized salvinia and duckweed floating on the top of the water. Yum! Those were my favorite plants to eat. I also spied guppies and mollies swimming in and out of the castle openings.

The fish all got along and the atmosphere was peaceful and tranquil. I realized I was the predator now and didn't want to scare them. If Terry fed me shrimp pellets and blood worms like she did in the classroom, I wouldn't have to eat any of my fellow tankmates. I wanted to keep peace in my new home. How terrible it was to live in fear of being eaten by a predator. I didn't want to cause the other animals to worry. I tried to communicate this to the fish and snails in my new habitat by quietly staying in the corner of the tank for a few hours while I watched them.

They stayed clear of me for a while and then the more curious ones swam over to check me out. Since I sat quietly and kept my claws still, they weren't scared and actually enjoyed smelling my exoskeleton and poking my tail. I felt like a big sister to them all. I wanted to be their friend and they sensed that knowledge.

Pirate Ship

I noticed that none of the fish were interested in the brown, craggy pirate ship that sat along the side of our tank. They preferred the colorful castle in the center of the tank that was bright and open. So I climbed on the pirate ship and claimed it as my home. I had never been on a pirate ship before. It was dark inside the ship and the little fish poked their heads in and out of the openings to see what I was doing.

I loved my life! My new environment was alive with activity but it was clean and non-threatening. How lucky for me!

Terry threw a few shrimp pellets into my tank and after I ate them, I lounged in the crow's nest of my pirate boat. I could see Terry in the distance, sitting on a chair and looking at a box that had moving pictures on it. She often watched the TV and I learned to look at it too. Sometimes there were shows about food and I watched how it was prepared in the kitchen. Seeing shrimp and lobster being cooked was a bit scary but I also watched animal doctor shows with Terry about making sick dogs, cats, horses, snakes, and guinea pigs better. Sometimes there were TV shows about people riding on horses chasing each other. When Terry's grandchildren came to visit, we watched colorful cartoons about talking dogs and other animals. There was even a cartoon about underwater sea creatures with a big yellow square that played with all of them. There was an octopus, a starfish and a snail in the show. However, I never saw any crayfish depicted on these shows. I wondered why not.

Every day, Terry came into our room and fed the fish in my tank some flakes. Then she looked for me and put something special in for me to eat. I liked the shrimp pellets and algae disks that Terry dropped gently into the water so that they floated right onto the deck of my pirate ship.

Once a week, she put some little bloodworms into my tank and the guppies and mollies tried to eat them as they fell to the bottom. I got whatever fell all the way down because the fish wanted to stay away from my claws as I caught the bloodworms. I enjoyed chasing the worms at the bottom of the tank for several hours. Some of them fell into the gravel on the bottom and then I felt around in the rocks with my antennae to discover their hiding places and ate them. A few times each week I found a rogue worm or two that evaded my capture and then I enjoyed a tasty treat mid week.

There was another tank next to mine with an extremely large gold-colored fish in it. Terry talked to him every night and called him Goldzilla. She seemed to love him. I knew that Terry liked me too. She often put her face up to the glass of my tank and smiled at me. She also put her finger to the glass and said, "Hello Gladys! Everything OK in there today?"

"Who is Gladys? Is that me? Wow! I guess Terry even gave me a name," I thought. I felt like an important crayfish since I never knew any other of my kind that had a name. I loved my life but always remembered the long journey it took to get to my forever home. Although I endured many things to finally end up in Terry's house, I ended up becoming the luckiest crayfish in the world.

Gladys's Vocabulary

Chapter 1:

Species - a type of living thing

Prey - an animal that is chased and eaten by another animal

Bait - good tasting food to try to catch an animal

Fate - karma or what happens to you in your lifetime

Bayou - a large swamp or bog

Segmented - having more than one part often times having many parts

Accommodate - to make room for something

Spinnerets - little legs on the underside of animals called crustaceans

Avian - birdlike

Bored - to push into something

Substrate - stuff on the bottom of a body of water

Predators - animals that chase and eat other animals

Oddities - things that are weird

Plethora - a lot of something

Appendages - any body part that sticks out like legs and arms

Rapid - quickly

Propelled - moved forward

Scurried - hurried to move

Decomposing - rotting

Chapter 2:

Dreary - rainy and cloudy

Waded - walked up to their knees

Cowering - hiding

Flailed - waved frantically

Rhythmically - to a steady beat

Severed - cut off

Nutrition - good, healthy food

Loner - being alone

Quarters - a place to live

Chapter 3:
Turrets - lookout places at the top of a castle
Recessed - pushed into
Stalks - stems or long sticks
Impediment - holding someone back
Exoskeleton - outside covering of a body
Protrusions - things that stick out
Torso - middle of a body
Streamlined - everything tucked in so you
    can move through water or air
Strain - hardship

Chapter 4:
Accurate - correct
Precision - detailed
Cephalothorax - part of a segmented body
Posed - questioned
Crustacean - an animal with a segmented
    body and an exoskeleton
Arthropod - an animal with many legs and
    a segmented body
Hollow - empty inside
Menacingly - with evil intentions
Marionette - puppet

Bulbous - fat and flabby

Ingests - eats

Mandibles - tooth like structures in some
   animals

Unison - as one

Chapter 5:

Putrid - smelly and rotting

Anticipation - waiting for an event to happen

Dejected - sad because something good
   didn't happen

Stance - a way to stand

Imploring - begging

Toxic - poison

Benthic - all the living things on and in the
   muddy bottom of water

Aquarium - a water environment

Trough - a big vat to hold water or food for a
   large animal

Gravity - the force pulling objects to the
   earth

Manmade - not natural, made by a person

Suction - pulling something into a place

Sensation - feeling

Apparatus - machine
Grate - opening like a screen
Debris - trash
Expelled - to push out
Spied - to see
Resourceful - being able to solve problems

Chapter 6:
Spewing - having a liquid spray out of
    something
Environment - a place to live
Tranquil - quiet and peaceful
Crow's nest - the top of a ship used as a
    lookout place
Depicted - shown
Rogue - wild, not doing as the others do
Evaded - got away from
Journey - a travel from one place to another
Endured - put up with

www.ingramcontent.com/pod-product-compliance
Lightning Source LLC
Chambersburg PA
CBHW071335200326
41520CB00013B/2999